Andy the Ant
Learns The Buzz On Bees

WRITTEN BY: Nancy Blackwell   ILLUSTRATED BY: Charles Ettinger

Andy the Ant Learns the Buzz on Bees

COPYRIGHT © 2013 by Nancy Blackwell

ISBN for paperback edition: 978-1-63073-001-7 ($14.99)
ISBN for hard cover edition: 978-1-63073-002-4 ($19.99)

Nancy Blackwell
296 Sunrise Harbor Road
Sunrise Beach, MO   65079

nbblackwell@hotmail.com

Published and Printed by:
Faithful Life Publishers
North Fort Myers, FL 33903

FaithfulLifePublishers.com
info@FaithfulLifePublishers.com

Photograph of the author by Jodi Hetherington
Photograph of the illustrator by Julie Lester

18   17   16   15   14   13   1   2   3   4   5

THANK YOU!

I would like to thank my husband, **Garry**, for helping me with this book. As a beekeeper for many years, he has taught me the passion that he has for bees. I can honestly say I have learned everything I know about beekeeping through him.

Dixie Phillips of Christian Writing Services.com for editing and the staff of Faithful Life Publishers for layout, design and publishing.

I would also like to thank **Grant Gillard**, past president of Missouri State Beekeepers Association, for his part in checking the manuscript for technical facts and content of the book.

Finally, I would like to thank the **National Honey Board** from Firestone, Colorado for allowing me to use their activity sheets in the back of this book.

"What a beautiful morning!" Andy smelled the sweet grass and felt the warm sun on his face. "Spring is finally here!"

As he climbed an old oak tree near his home, the little ant's body vibrated with excitement. When he reached the highest branch, he peered down on the grassy meadow and something caught his eye. "Hmmmmm. What could that be?"

Several white wooden boxes were stacked side by side. Andy darted down the tree and tiptoed toward them. Honeybees flew in and out of the boxes. Andy crawled up on the side of the box and poked his head around the corner to watch. He was startled when an angry bee rushed toward him. He lost his footing and fell to the ground. Thump!

The bee dove at Andy and scolded, **"Stay away!"**

Andy ducked under the box and hunkered down.

Another bee buzzed toward Andy. "Hello, little ant. What are you doing here?"

"Just lookin' around." Andy looked over his shoulder to make sure it was safe.

"My name is Beatrice. What is yours?"

"I'm Andy the Ant."

"I am sorry my sister frightened you, but she is a guard bee. Her job is to protect the **beehive** and its honey from danger." Beatrice smiled at Andy.

"Really?" Andy let out an exaggerated sigh. "Well, she sure scared me".

A clattering sound interrupted the new friends. Beatrice pointed to an old pickup truck. "Sounds like Ben and his father, who are our beekeepers, have arrived." Andy wrinkled his nose. "What is a beekeeper?"

"Beekeepers raise bees and care for them. They also collect the extra honey we make for people to use." Beatrice's smile turned upside down. "Sometimes we get sick and the beekeepers have to give us medicine to help us get well."

"Really?" Andy's antennas stood at attention.

"Yep!" Beatrice nodded. "Beekeepers have to work really hard. They even watch out for other insect pests that can get into the hive and destroy it."

"What are they doing?

Why are they putting on those strange helmets which have veils covering their faces?" Andy watched the beekeepers throw on a white suit and long gloves over their clothes.

"Bees can see dark colors better, so the beekeepers wear white outfits to protect them from getting stung. Don't they look funny?" Beatrice covered her mouth and giggled.

"They sure do!" Andy chuckled.

"Beekeepers also need special tools to help them work around the hives. Let's see . . ." Beatrice leaned back, thinking. "They need the **hive tool**, the **smoker**, and the **bee brush**."

"Wow! You are one smart bee!"

Beatrice buzzed loudly. "Andy, would you like to see what's inside our hive?"

"I would, but will the bees hurt me?"

"They better not!" Beatrice pursed her lips. "Before we go inside, would you like me to tell you more about us bees?"

Andy cleared his throat. "I sure would!"

"Did you know a honeybee is an insect? They live in colonies just like you do."

"Really?"

"Yep. There are three kinds of bees that make up a hive: **drones, the queen,** and **workers**." Beatrice curtsied quickly and continued. "Drones are male bees. They mate with the queen to make babies. The Queen's job is very important. She lays eggs in the hive. She can lay as many as 2,000 eggs a day. There is only one queen in a hive and she is bigger than the rest of the bees. Can you imagine, we are all her children? Workers are female bees that do most of the work inside the hive. They clean the hive, make wax and honey, care for the baby bees, and feed the queen. "

"Wow! That's amazing!" Andy squeezed his eyes closed and sighed. "I understand why you call her the queen bee."

Drone   Queen   Worker

"Follow me, Andy." Beatrice waved her hand. "I will show you what is inside."

Andy followed closely behind Beatrice. "It's kind of dark in here."

"You are safe with me." Beatrice's eyes twinkled. "The hive has several boxes filled with **frames**. This is the entrance area. The brood chamber is in the bottom box, where the queen lays her eggs and the young bees are born. The next box is the food chamber, where the bees store honey for their food. The top box, way up there, is called the honey super. That is where the bees store their extra honey."

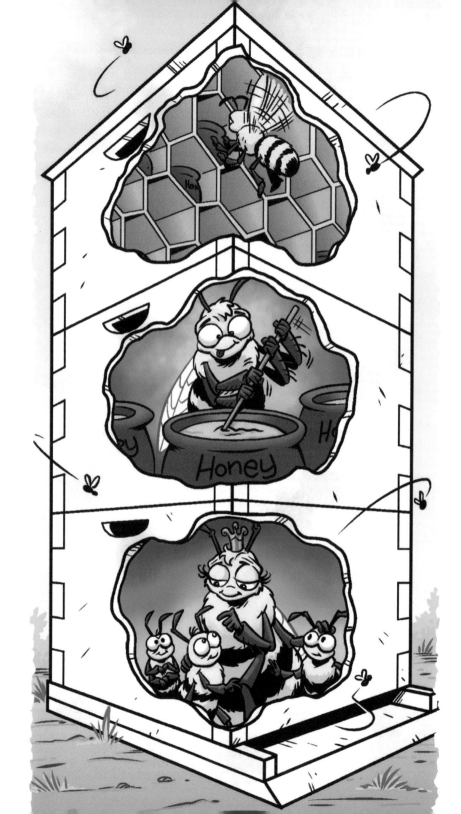

While Beatrice was explaining the beehive to Andy, Ben opened up the lid with his hive tool. "Oh my, there's an ant in here!"

Beatrice quickly spoke up. "Ben, this is Andy the Ant. He is my new friend and I am showing him the hive."

Ben winked at Andy. "Nice to meet you."

"You too." Andy bowed.

"What are you looking for, Ben?" Beatrice flew up by Ben's ear.

"I need to see if the bees are collecting nectar and making honey."

"Could I taste some honey?" Andy's voice was filled with excitement.

"Sure!" Ben put a dab of honey on his finger and Andy licked it.

"Delicious! It is very sweet." Andy licked the corners of his mouth. "Could I take some home with me, so the rest of the ants could have a taste?"

"You sure can!" Ben replied.

Beatrice flew back to Andy's side. "Before you go, Andy, I wanted to share with you how bees make honey."

Andy shifted his attention back to Beatrice. "OK."

"In the spring the bees fly out of their hives at first light. The workers or **forager bees** zoom from flower to flower looking for **nectar** and **pollen**. When a bee finds a flower, she begins sucking out the sweet nectar using her long tongue and stores the nectar in her **honey stomach**. At each flower she crawls around inside the flower and gets covered by a yellow dust called pollen. She then packs the pollen into pollen baskets on her back legs. When she lands in another flower, pollen falls off her body into a flowering plant to make seeds and new plants. This is called pollination."

"Look, Beatrice!" Andy's eyes grew wider. "What are those bees doing over there?"

"They are dancing to show the other bees where the flowers are." A grin lifted the corners of Beatrice's mouth. "They do a **round dance** to show that nectar is nearby and a **waggle dance** when the nectar is far away."

"Golly! I never knew bees communicated by dancing. Do you suppose they know how to jitterbug?" Andy teased.

"Nope, but they sure know how to dance—as if they have ants in their pants!" Beatrice fluttered her eyes.

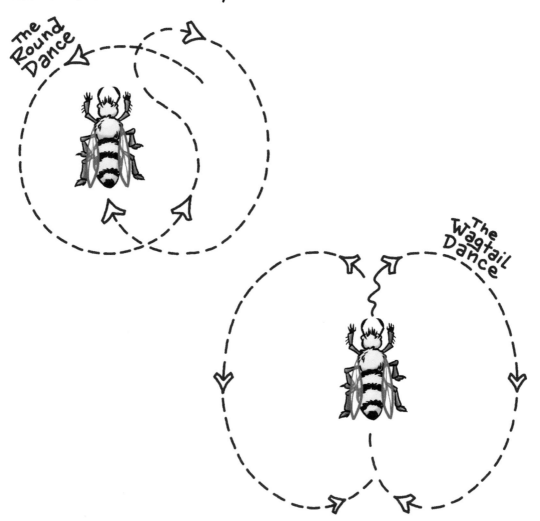

Andy laughed, his dark eyes sparkling. "Where do the bees go from here?"

"The bees fly back to their hive and store the nectar in the frames to make honey." Beatrice waved her hand in the direction of the hive.

"Thanks for telling me how bees make honey." Andy headed toward the door. "I have to go home now. I'll see you later."

"Goodbye, Andy."

"Goodbye, Beatrice."

Early the next morning, Andy approached the bee yard and could not believe his eyes. A mother bear and her cub were knocking over the hives and ripping honey from the frames. Andy knew he had to do something to protect Beatrice and the hive. "Stop it! You should be ashamed of yourselves for knocking over the hives and hurting the bees!"

"Go away, little ant, and mind your own business!" growled the bears.

Out of nowhere, Ben and his father appeared.

Andy shouted as loudly as he could, **"BEAR ATTACK!"**

Immediately the beekeepers galloped toward the bears, shouting and waving their hands. The bears scurried into the woods. Ben and his father approached the damaged hives and started to put them back together.

"Oh, no! The bears destroyed Beatrice's hive!" Andy looked around for his new friend. His heart thumped as he yelled, "Beatrice! Beatrice!" Andy heard a soft whimper. He spotted a bee trapped under a piece of wood. "Beatrice, is that you?"

"I can't move!" Beatrice cried.

Ben appeared on the scene. "Father, come quick! Some of the bees are trapped."

As Father ran over, Andy pleaded with Ben, "Please help Beatrice. She's caught and can't get out."

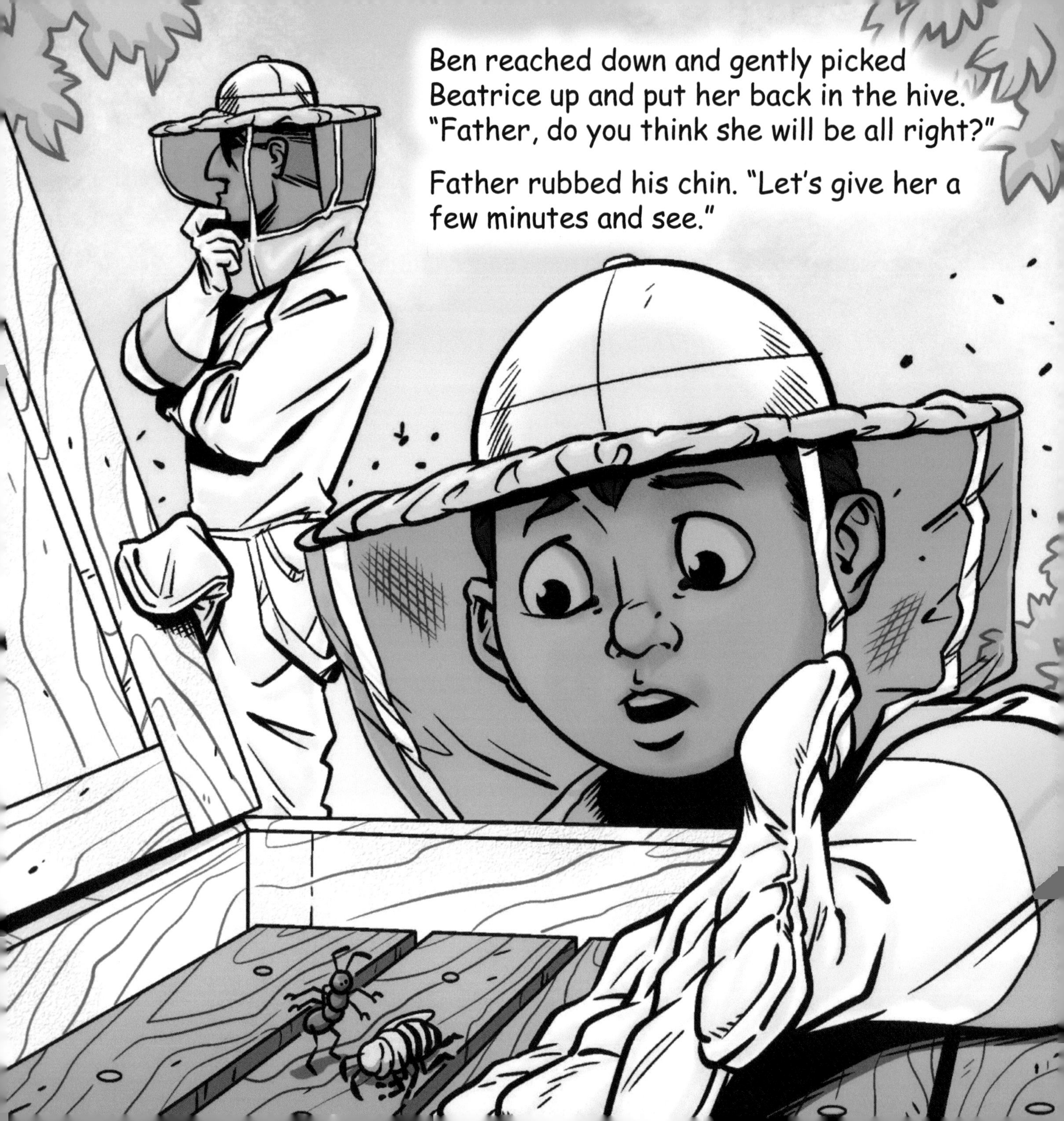

Ben reached down and gently picked Beatrice up and put her back in the hive. "Father, do you think she will be all right?"

Father rubbed his chin. "Let's give her a few minutes and see."

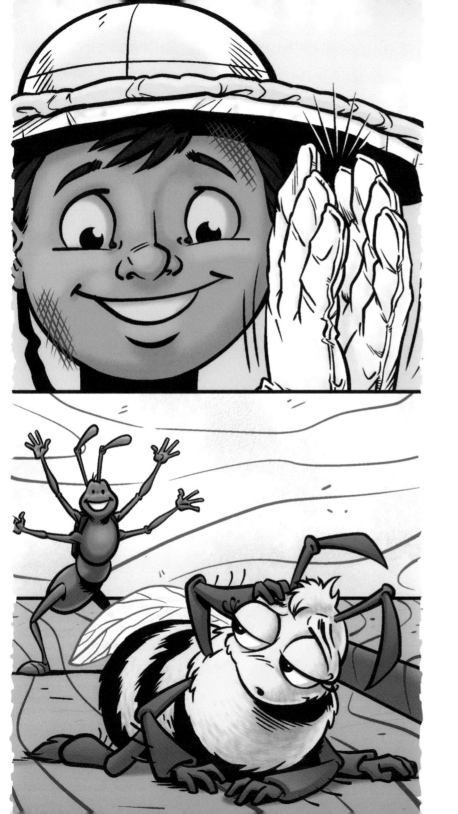

"Look! She's starting to move!" Ben clapped his hands. "I hope her family will take care of her."

Father squeezed his son's shoulders. "Putting Beatrice back into the hive was a smart thing to do, Ben! You saved her life."

"Hooray for the bees!" Ben let out a whoop. "Look how they are caring for Beatrice."

Andy crawled closer to Ben and asked, "Isn't there something you and your father can do to keep the bears away from the hives?"

"Father and I were just talking about that. We plan to put an electric fence around the hives!" Ben gave Andy a high-five.

"That's a great idea!" Andy whistled. "Then they will be safe from the bears."

Later in the day Andy heard a buzzing sound coming from the tree. He looked up and saw a big ball of bees hanging on a branch. "Beatrice, are you up there?"

"Yes, we are swarming and looking for a new home. Right now the scout bees are searching for a good spot for a new hive." Beatrice's wings beat feverishly. "Sometimes when our hives are disturbed, we need to find another home. God knows what is best for all of us. I hope I will see you again someday. Goodbye, Andy." Suddenly the bees flew away in a big cloud.

The next day Andy approached the bee yard. Ben and his father were studying the hives. "What are you doing, Ben?"

"We are heading over to our **honey house** to do some work. Want to come?"

"Sure!"

Ben picked Andy up, put him on his shirt collar, and headed to the honey house.

Once they arrived, Ben explained, "Father and I have already harvested the frames of honey by removing them from the hives. Then Father will cut the **wax caps** off with a hot knife and place the frames inside the extractor."

"What's an extractor?" Andy asked.

"It's a big machine that spins the frames around until the honey is removed from the frames. The beekeeper drains the honey into big storage buckets and then pours the honey into bottles to sell," Ben explained.

Andy and Ben watched as Father finished up his work. "Well, we're all done for today," he announced. Ben turned to Andy, "It's time for me to go; but if you like, I'll be glad to take you home."

"Thanks, Ben. I can hardly wait to come back again and learn some more about beekeeping."

Andy snuggled down onto Ben's collar for the ride home. As they left the bee yard, he thought about Beatrice and wondered where his friend had found a new home.

Andy learned that bees are an important part of the world. Everyone needs to appreciate and protect them, so they can continue producing for themselves, pollinating plants, and providing honey for people to eat.

For thousands of years the honeybee has been a very important insect to nature and to man. After reading this book, I hope you "bee"lieve bees are not only amazing, but also hardworking creatures. The expression "busy as a bee" means a bee is active all the time. Bees work hard and do their best throughout their lifetime. We can learn from the bee's example about the importance of working together and being a hard worker. Then we can "bee" the best that we can "bee."

Un "bee"lievable Facts about Honeybees

- When the hive becomes too hot, the bees use their wings to fan and cool the hive.

- Honeybees never sleep.

- Each bee will produce a drop of honey in its lifetime.

- The beekeeper must be careful not to take too much honey from the hive or the bees will starve during the winter.

- Sometimes the beekeeper has to feed the bees sugar syrup.

- Yellow jackets feed honey bees to their young.

- Only female bees have stingers.

- Worker bees die shortly after stinging.

- Queen bees can live as long as four to five years.

- Queen bees eat a special food called royal jelly.

- A worker bee can fly up to 15 miles per hour.

- A hive can produce up to 500 pounds of honey a year, depending how much the bees bring in.

- The baby eggs take 21 days to become bees.

- The bees go through these stages: egg, larva, pupa, and adult.

- Making one pound of honey is the lifetime work of approximately 300 bees. The bees must visit two million flowers and fly over 55,000 miles.

- A hive can have as many as 50,000 to 60,000 bees during the spring and summer months.

- Other Products from Beehives:

  1. Bee pollen is used for protein supplement.

  2. Royal jelly can be used for nutrition and is sold in health food stores.

  3. Bee wax is used to make candles, crayons, lipstick, lotion, and polishes for furniture and cars.

Glossary

Beehive   A box or series of boxes that house thousands of bees.

Hive tool   A metal tool used to help open the hive, remove the frames, and scrape the surfaces clean.

Smoker   A metal container that burns dry leaves or paper to make smoke. The smoke is then blown into the hive to calm the bees.

Bee brush   A tool used to brush bees from the frames.

Frames   Four pieces of wood that form a rectangle designed to hold the honeycomb.

Forager bee   A worker bee that gathers food and defends the hive.

Nectar   A sweet liquid found in many flowers.

Pollen   A yellow powdery substance found inside flowers.

Honey stomach   A special stomach to hold a large amount of liquid to store nectar.

Round dance   Bees dance in circles, switching directions.

Waggle dance   Bees dance in a figure-eight pattern.

Swarming   Worker bees, drones, and queens leave the original hive to build a new bee colony.

Honey house   A building used for extracting honey, packaging honey, and storing other equipment.

Wax cap   A wax cover that bees place over a cell of honey.

ACTIVITY PAGES

Word Search

Circle the following honey bee words in the puzzle below:

BEEKEEPER, BEESWAX, BUZZ, CLOVER, COLONY, COMB, DRONE, FLOWER, FORAGE, GARDEN, HEXAGON, HIVE, HONEY, INSECT, NECTAR, PETAL, POLLEN, POLLINATION, QUEEN, WORKER

```
B N P O L L I N A T I O N
J E E Y W R L T V Y F N J
V X E L T L A C L O V E R
E D U S L V T O C R T K T
K R W A W O E L D K Q R C
H O N E Y A P O Q U F O E
I N E Y Y Y X N M L M B S
V E C N G Q L Y O B Q E N
E Y T O Y L E W U A D E I
Q K A G M N E G R G O K Q
B U R A B R Z K E W E E W
U W E X X K U G K R A E C
Z O E E S N E D R A G P L
Z S K H N C X J O F F E Q
F O R A G E L D W H U R U
```

Beekeepers

A Beekeeper's Equipment

Using the list of hive elements and beekeeper's tools, label the following.

Bottom board
Coveralls
Frame
Gloves
Hive body or
  brood chamber
Helmet

Honey super
Inner cover
Outer cover
Smoker
Veil

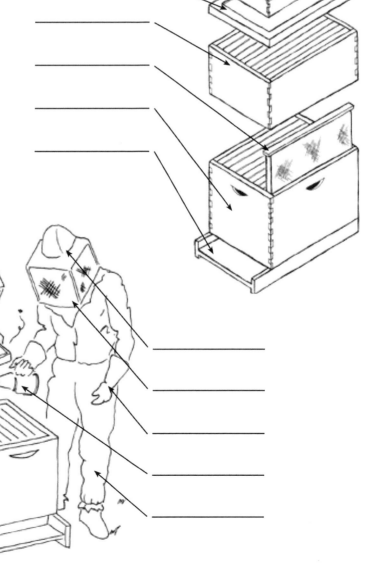

Honey

Kid's Recipes

Ovens are hot, knives are sharp, blenders are fast and microwaves can be tricky. The point? Kids need adult helpers in the kitchen. Adult supervisors can read recipes with kids so the directions are clear. Adult helpers can also assist with kitchen utensils, machines and appliances. Play it safe and prepare these honey recipes together.

Banana Pops

1-1/3	cups topping, such as ground toasted almonds, toasted coconut, candy sprinkles or graham cracker crumbs
4	bananas, peeled
8	wooden craft sticks
1/2	cup honey

Spread toppings of your choice on a plate or plates. Cut bananas in half crosswise. Insert a craft stick into each cut end. To assemble, hold 1 banana over plate or waxed paper to catch drips. Spoon about 1 tablespoon honey over banana, rotating and smoothing honey with back of spoon to coat all sides. (Or squeeze honey from a plastic honey bear container and smooth out with spoon.) Roll banana in topping of choice until coated on all sides, pressing with fingertips to help topping adhere. Place pops on waxed paper-lined cookie sheet. Repeat with remaining bananas, honey and toppings. Serve at once. **Makes 8 servings.**

Bee Nutty Choco-Chip Cookies

1/2	cup honey
1/2	cup peanut butter
1/2	cup butter or margarine
1/4	cup packed brown sugar
1	egg
1-1/2	teaspoons vanilla
2	cups flour
1/2	teaspoon each baking soda and salt
1	cup chocolate morsels
1/2	cup coarsely chopped roasted peanuts

Combine honey, peanut butter, butter and brown sugar in a large bowl; beat until light and fluffy. Add egg and vanilla; mix thoroughly. Combine flour, soda and salt; mix well. Stir into peanut butter mixture. Stir in chocolate morsels and peanuts. Using 1/4 cup for each cookie, drop onto ungreased cookie sheet; flatten slightly. Bake at 350°F 8 to 10 minutes or until lightly browned. Remove to rack and cool. **Makes 16 cookies.**

Honey

Kid's Recipes continued

Honey Lemonade with Frozen Fruit Cubes

1-1/2 cups lemon juice
3/4 cup honey
9 cups water
28 pieces assorted fruit

In pitcher, whisk lemon juice with honey to dissolve. Whisk in water. Place 1 piece of fruit in each compartment of 2 ice trays. Fill each compartment with honey lemonade. Freeze until firm. To serve, divide frozen fruit cubes between glasses; fill with remaining lemonade. Serve in tall glasses. **Makes 6 servings.**

Honey Care to Take a Dip

1 pint (16 oz.) lowfat plain yogurt
1/4 cup honey
2 Tablespoons orange juice
1/2 teaspoon grated orange peel
Assorted fruits for dipping such as sliced apples, pears and strawberries

Combine yogurt in a small bowl with honey, orange juice and orange peel; mix well. Serve with sliced fruit. **Makes 2-1/4 cups.**

Honey Crispies

1/2 cup powdered sugar
1/2 cup honey
1/2 cup peanut butter
1-1/2 cups crispy rice cereal
1/2 cup raisins
1/2 cup chocolate or multicolored candy sprinkles

Place a sheet of waxed paper on a cookie sheet so cookies won't stick. Combine powdered sugar, honey and peanut butter in a medium bowl. Stir until mixed well. Stir in cereal and raisins. Using hands, shape mixture into 1-inch balls. Roll balls in sprinkles and place on a cookie sheet. Refrigerate for 1 hour. Cookies should feel firm when touched. Serve right away or place in tightly covered container and store in refrigerator. **Makes about 30 cookies.**

ABOUT THE AUTHOR:

Nancy Blackwell taught kindergarten and elementary school for fifteen years. Her teaching experience gave her an intimate understanding of how children see the world. She still loves sharing stories with schoolchildren, as well as her own grandchildren. Nancy wants to provide youngsters with positive, moral stories written from a Christian perspective. Because she loves the outdoors, her favorite stories are those that teach about nature.

Join Andy the Ant as he treks through an exciting adventure inside a beehive with his new friend, Beatrice the Bee. The author weaves a fun and educational story about beekeeping. Although this charming story is designed for elementary children, adult readers may learn a thing or two as well! The book is loaded with fun-filled information about the duties of a beekeeper, the various jobs bees have, how bees make honey, why bees dance, and more.

Andy the Ant Learns the Buzz on Bees provides highlighted vocabulary words throughout the story. Activities in the back of the book include some tasty honey recipes for kids to try at home.

Parents, children, and teachers will be "abuzz" with all the un"bee"lievable facts of this charming and informative story. It is an excellent science resource for the little scientists in your life.

ABOUT THE ILLUSTRATOR:

Charles Ettinger is a freelance artist and has spent the majority of his life in Memphis, Tennessee. He has a background in commercial art, graphic design, comic books, promotional work, and children illustrations. Charles will delight everyone as the characters come alive through his illustrations.

CPSIA information can be obtained
at www.ICGtesting.com
Printed in the USA
395776LV00004B/10